AF812153

Cyber Shield

Defending Your Digital Life

Tyler L. Strong

Other Books by Tyler L Strong: **Breaking Into It: Your Roadmap to Success in the Tech Industry**

Cyber Shield: Defending Your Digital Life Copyright © 2024 by Tyler L Strong. All rights reserved. No part of this publication may be reproduced, distributed, or transmitted in any form or by any means, including photocopying, recording, or other electronic or mechanical methods, without the prior written permission of the author, except in the case of brief quotations embodied in critical reviews and certain other noncommercial uses permitted by copyright law. Requests for permission to use material from this book (other than for review purposes) should be sent to contact@unitethegrid.com. At the time of publishing, Tyler L Strong serves as the founder of The GRID.

ISBN's

Paperback - 978-1-964031-04-0

eBook - 978-1-964031-03-3

Audio - 978-1-964031-05-7

Disclaimer: This book is for informational and educational purposes only. The characters and events in case studies are entirely fictional; any resemblance to actual persons, living or dead, or actual events, is coincidental. While efforts have been made to ensure accuracy, the author and publisher do not assume responsibility for errors, omissions, or different interpretations of the subject matter. The information should not be construed as professional advice, and any actions taken based on this book's content are at the reader's discretion. The author and publisher disclaim liability and responsibility to any person or entity with respect to any loss or damage caused or alleged to be caused directly or indirectly by the use of the information in this book.

Table of Contents

Chapter 1: Understanding the Digital Threat Landscape ________ 8

Introduction to Cybersecurity ________ 8

The Mechanics of Cyber Threats ________ 10

The Impact of Cyber Threats on Society ________ 12

Chapter 2: Password Hygiene and Management ________ 15

The Importance of Strong Passwords ________ 15

The Role and Benefits of Password Managers ________ 16

Two-Factor Authentication (2FA) ________ 17

Chapter 3: Safe Web Browsing Practices ________ 20

Secure Browsing Habits ________ 20

Privacy Tools and Extensions ________ 21

Chapter 4: Phishing Attacks ________ 24

Understanding Phishing ________ 24

Best Practices for Verification ________ 26

Chapter 5: Social Media Security ________ 29

Understanding Social Media Risks ________ 29

Being Mindful on Social-Media ________ 30

Chapter 6: Protecting Sensitive Information ________ 33

Understanding Encryption ________ 33

Broadening Your Digital Defense: Practical Security Measures _ 35

Chapter 7: Staying Informed and Up-to-Date ________ 39

The Importance of Continuous Learning ________ 39

Educational Resources and Courses ________ 40

Conclusion ________ 43

Appendix ________ 44

Introduction

In today's interconnected world, the internet has become an integral part of our daily lives, offering unparalleled convenience, access to information, and means of communication. However, this digital revolution comes with its own set of challenges, primarily the increasing threat to our personal information and privacy. "Cyber Shield: Defending Your Digital Life" is your essential guide to understanding these threats and arming yourself with the knowledge and tools necessary to safeguard your digital presence.

This book is designed for individuals at all levels of digital literacy, offering straightforward, actionable advice to protect against cyber threats. From basic internet users to seasoned tech professionals, everyone will find valuable insights and practical tips within these pages. Our journey will take you through the landscape of digital threats, into the world of password security, safe web browsing practices, and beyond. With this guide, you'll learn how to defend against cyber-attacks and how to build a proactive mindset towards digital security, ensuring that you and your information remain safe in an ever-evolving digital world. If you find this book helpful, I encourage you to return to the platform where you obtained the book to provide a 5-star review. This will help others find the book, allowing it to reach a wider audience.

Chapter 1: Understanding the Digital Threat Landscape

Welcome to the first step in fortifying your digital life. This chapter lays the groundwork for our journey into cybersecurity by introducing you to the digital threat landscape. In an age where our lives are increasingly online, understanding these threats is the first line of defense in protecting your personal information, privacy, and digital wellbeing. We'll explore the importance of cybersecurity, unravel the types of cyber threats lurking in the digital shadows, and share real-world examples to illustrate the impact these threats can have on your life. By the end of this chapter, you'll have a solid understanding of the digital dangers that exist and why it's crucial to take proactive steps towards securing your online presence.

Introduction to Cybersecurity

Cybersecurity is the practice of protecting systems, networks, and programs from digital attacks. These attacks are usually aimed at accessing, changing, or destroying sensitive information; extorting money from users; or interrupting normal business processes. Implementing effective cybersecurity measures is particularly challenging today because there are more devices than people, and attackers are becoming more innovative.

The relevance of cybersecurity cannot be overstated. In our digital age, virtually everyone is connected in some way to the digital world, whether through email, social media, online banking, or shopping. This interconnectedness, while beneficial, also opens up numerous avenues for cybercriminals to exploit. Personal information, such as social security numbers, bank account details, and email addresses, are valuable commodities on the dark web. A breach of this information can lead to identity theft, financial loss, and a significant invasion of privacy.

Glossary of Important Terms

Before we get into the complexities of the digital threat landscape, let's clarify some key terms you'll encounter throughout this guide. Understanding these concepts will help in grasping the full scope of cybersecurity challenges and solutions discussed in the following chapters.

- **Malware:** Short for malicious software, it refers to any program or file designed to harm or exploit any programmable device, service, or network. Malware includes viruses, worms, Trojans, and ransomware.
- **Ransomware:** A type of malware that encrypts the victim's files, making them inaccessible, and demands a ransom payment to decrypt them.
- **Phishing:** A cyber-attack that uses disguised email as a weapon. The goal is to trick the email recipient into believing that the message is something they want or need — a request from their bank, for instance, or a note from someone in their company — and to click a link or download an attachment.
- **Social Engineering:** A manipulation technique that exploits human error to gain private information, access, or valuables. In cybersecurity, social engineering is a tactic used to deceive users into making security mistakes or giving away sensitive information.
- **IoT (Internet of Things):** A network of physical objects ("things") that are embedded with sensors, software, and other technologies for the purpose of connecting and exchanging data with other devices and systems over the internet. These devices range from ordinary household items (smartphone, laptop, tablet, smart watch, etc.) to sophisticated industrial tools.
- **Encryption:** The process of converting information or data into a code, especially to prevent unauthorized access. Encryption is a critical tool for protecting data's confidentiality and integrity, both in transit and at rest.
- **VPN (Virtual Private Network):** A service that encrypts your internet connection and hides your online identity. By creating a secure, encrypted tunnel for your data, a VPN ensures that your internet activity is not monitored or recorded.
- **Two-Factor Authentication (2FA):** An extra layer of security used to ensure that people trying to gain access to an online account are who they say they are. First, a user will enter their username and a password. Then,

instead of immediately gaining access, they will be required to provide another piece of information.

The importance of cybersecurity grows as the number of internet users continues to rise, and our reliance on digital services becomes ever more integral to our daily lives. By understanding the threats and recognizing the signs of an attack, individuals can significantly reduce their vulnerability to cybercrimes.

Understanding Malware Through Analogy

Imagine if your computer caught a cold, just like you might in the winter months. This "cold" is what we call malware in the digital world—a type of software designed to disrupt, damage, or gain unauthorized access to your computer system. Just as a cold virus spreads from person to person, malware spreads from computer to computer, infecting systems, stealing information, or causing other harm. There are many types of malware, including viruses, which replicate themselves; worms, which spread across networks; and Trojans, which disguise themselves as harmless software. Protecting your devices against malware is akin to practicing good hygiene to prevent the flu—necessary and ongoing.

The Mechanics of Cyber Threats

Cyber threats exploit vulnerabilities in software, human psychology, and system configurations to gain unauthorized access or cause harm. For example, malware might exploit a security flaw in an outdated software program to infect a computer. Once inside, it can steal information, corrupt files, or even take control of the system for malicious purposes.

Ransomware operates by encrypting the victim's data, making it inaccessible until a ransom is paid. This type of malware often enters systems through phishing emails or exploiting software vulnerabilities. Phishing attacks and social engineering exploit human psychology. By masquerading as a trustworthy entity, attackers deceive individuals into providing sensitive information, such as passwords or financial details. These attacks often rely on urgent language and plausible requests to lower the victim's guard.

Preventive Measures Overview: To defend against these various threats, adopt a proactive security posture:

- Regularly update software and operating systems to close security loopholes. This includes devices like smart-phones, laptops, and tablets.
- Use strong, unique passwords for different accounts and consider using a password manager.
- Exercise caution with emails, especially those urging immediate action or requesting personal information. Verify the sender's credibility before responding or clicking on links.
- Utilize antivirus software to protect your systems

Popular Antivirus Tools:

1. **Bitdefender**: Offers comprehensive protection with multi-layered anti-malware, anti-phishing, and anti-fraud features, along with minimal impact on system perfor-mance.
2. **Norton 360**: Provides not only antivirus but also a VPN, password manager, and cloud backup, making it a versatile security suite.
3. **McAfee Total Protection**: Includes antivirus, identity theft protection, and secure browsing features, suitable for individuals and families.

The Importance of Software Updates

One of the simplest yet most effective steps you can take to protect your digital life is to keep your software updated. Think of software updates as booster shots for your devices. Just as a flu vaccine is updated each year to combat new strains of the virus, software updates often include patches for security vulnerabilities that have been discovered since the last update. By keeping your operating system, applications, and antivirus software up to date, you're ensuring your devices have the latest defenses against mal-ware and other cyber threats.

The Impact of Cyber Threats on Society

Cyber threats have far-reaching effects beyond individual inconvenience or financial loss. Large-scale data breaches can compromise the personal information of millions, leading to wide-spread identity theft and financial fraud. On a national level, cyber-attacks can target critical infrastructure, posing threats to public safety and national security.

Emerging Threats

Emerging threats like deepfakes—synthetic media in which a person's likeness is replaced with someone else's—pose new challenges for information authenticity. IoT vulnerabilities expose personal data and the functionality of connected devices, from smart homes to city infrastructures. AI-driven phishing uses machine learning to create highly convincing fake messages, making traditional detection methods less effective.

Phishing Case Study: In this scenario, a company's CFO receives an email from the CEO (or so it seems), requesting an urgent wire transfer to finalize a confidential acquisition. The email contains the company's letterhead, the CEO's signature, and specific details about the company's financial transactions, making it appear legitimate. However, the email is a sophisticated phishing attempt. The attacker researched the company and its executives, using this information to craft a convincing fake email. The CFO, pressured by the supposed urgency and authenticity of the request, initiates the transfer, resulting in significant financial loss.

Preventive Measures:

- Verify unusual requests through multiple channels, e.g., a direct phone call to the supposed sender.
- Implement company-wide protocols for financial transact-tions that require dual authentication.

Further Reading Links

For readers seeking to dive deeper into cybersecurity topics, consider exploring the following resources:

- **StaySafeOnline** (https://staysafeonline.org): Provides cybersecurity awareness and education, including tips on protecting personal information.
- **Cybersecurity and Infrastructure Security Agency (CISA)** (https://www.cisa.gov): Offers resources and alerts on current cybersecurity threats and vulnerabilities.
- **Krebs on Security** (https://krebsonsecurity.com): A blog dedicated to investigative stories on cybercrime and cybersecurity.

As we wrap up this chapter, it's clear that the digital world, while offering incredible opportunities for connection and learning, also presents significant risks to our personal information and privacy. Understanding the variety of threats that exist online is the first critical step in protecting yourself. Armed with this knowledge, you can start to build a more secure online presence for yourself and your loved ones. Remember, cybersecurity is about being aware and taking small, consistent steps to safeguard your digital footprint. Let's move forward with confidence, knowing that we have the tools and knowledge to navigate the online world more safely.

Reflect and Apply Questions

1. **Reflect on Your Current Practices:**
    - Have you ever received an email that made you pause and wonder if it was legitimate? What were the signs that alerted you?
    - Think about your current passwords. Are they strong and unique across different accounts? If not, what's one step you could take today to improve your password security?
2. **Consider Your Digital Footprint:**
    - Reflect on the amount of personal information you share online, especially on social media. Can you identify any information that might be better kept private?
    - When was the last time you checked the privacy settings on your social media accounts? Are there adjustments you could make to better protect your information?
3. **Evaluate Your Software Security:**
    - When did you last update the operating system on your computer or smartphone? Do you have automatic updates enabled, or do you prefer to update manually?
    - Do you use any form of antivirus or anti-malware software? How regularly do you check that it's updated and functioning correctly?

Chapter 2: Password Hygiene and Management

In the digital age, our lives are secured by a series of secret keys: our passwords. They guard the gates to our personal, financial, and work-related worlds. Yet, despite their critical role, password management is often overlooked, leaving many vulnerable to cyber-attacks. This chapter explores the art of creating strong passwords, avoiding common pitfalls, and leveraging tools like password managers to fortify your digital security.

The Importance of Strong Passwords

A strong password acts as a robust barrier between cyber-criminals and your personal information. But what makes a password strong? It's a combination of length, complexity, and unpredictability. A strong password should be at least 12 characters long, mixing letters, numbers, and special characters to create a sequence that's both unique and challenging for attackers to guess.

Weak passwords, on the other hand, are the low-hanging fruit for cybercriminals. They often use automated tools that can guess thousands of passwords per second, quickly cracking simple or common passwords. The consequences of a weak password can range from unauthorized access to your social media accounts to more severe outcomes like identity theft or financial loss.

Common Password Pitfalls

One of the most significant risks in password security is the use of predictable passwords. Common examples include "password-123," "123456," and the use of easily guessable information like birthdates or pet names. These passwords are easily breached, making your accounts low-effort targets for hackers.

Another widespread issue is password reuse across multiple accounts. This practice means that if one account is compromised, others are at risk too. A study by the National Cyber Security Centre revealed that millions of users are using passwords as simple as "123456" across multiple platforms, exponentially increasing their vulnerability.

Creating and Remembering Strong Passwords

Creating a strong password doesn't have to be a memory test. One effective method is to use a passphrase—a series of random words or a sentence that you can remember but is hard for others to guess. For instance, "BlueMonkeyMountainBicycle" is long, in-cludes a mix of concepts, and is more secure than a simple word or phrase. But how do you remember these complex passwords, especially when you have a different password for each account? Here are a few tips:

- Use mnemonic devices to remember complex passwords or passphrases.
- Regularly review and rehearse your passwords to keep them fresh in your memory.
- Securely store a written copy in a locked, private location if absolutely necessary.

The Role and Benefits of Password Managers

A password manager is a tool that stores and manages your passwords in a secure, encrypted database. You only need to remember one strong master password to access your entire password library. Password managers can also generate strong, random passwords for each of your accounts, ensuring that each password is unique and secure. Benefits of using a password manager include:

- **Convenience:** Auto-fill your login information on websites and apps.
- **Security:** Store passwords in an encrypted format, making them nearly impossible for hackers to access.
- **Peace of Mind:** Never worry about forgetting a password again, or the security risks of writing them down or reusing them.

Common concerns about password managers often revolve around the safety of storing all passwords in one place. However, reputable password managers protect your data with advanced encryption techniques. Even if a password manager's database were somehow breached, the encryption would keep your passwords safe from decryption. Popular Password Managers include:

LastPass stands out for its user-friendly interface and variety of features, including secure password storage, autofill, password generation, and secure notes for sensitive information. It's available on multiple platforms, ensuring seamless access across devices.

Bitwarden is noted for its open-source nature, offering transparency in its security practices. It provides similar features to LastPass, such as secure storage and password generation, but with the added benefit of user auditability, making it a favorite for those who prioritize transparency and control.

1Password is known for its strong focus on ease of use and design, providing features like secure password storage, document storage, and travel mode for added security. It's particularly favored by families and teams with its sharing capabilities.

Dashlane offers a robust password management solution with added benefits like dark web monitoring and a VPN for secure browsing. Its user interface and security dashboard, which assesses password health, make it a popular choice among users looking for comprehensive security tools.

Two-Factor Authentication (2FA)

Two-Factor Authentication (2FA) significantly enhances your security by requiring a second form of verification beyond just your password. This could be a text message with a code, a biometric factor like a fingerprint, or an authentication app. By enabling 2FA on your accounts, you make it much harder for attackers to gain unauthorized access, even if they somehow discover your password. Popular MFA (Multi-Factor Authentication) providers include:

- **Duo Security**: Known for its wide range of integrations with other platforms and user-friendly experience.
- **Google Authenticator:** Offers a straightforward, cost-effective solution for generating time-based one-time passwords.
- **Authy:** Stands out for its ability to sync across multiple devices, providing a seamless experience

for users managing MFA tokens on more than one device.

Each of these providers offers a unique set of benefits, such as ease of use, integration capabilities, and enhanced security features to protect user accounts across various services.

Best Practices for Password Security

To maintain robust password security, follow these best practices:

- Regularly update your passwords, especially for sensitive accounts.
- Use a unique password for each account to prevent a single breach from compromising all your accounts.
- Be cautious about sharing passwords, even with trusted individuals or platforms.

Protect Your Lock Screen: One often overlooked form of protection is the lock screen of your smart device. For convenience, many opt to have applications auto-log in. This renders a complex password or password manager useless if a malicious individual is able to not only get access to your phone, but access it with a simple swipe. It is important to set up facial recognition, fingerprint scanning, a pin, or any other type of lock screen security. While it can feel like a headache having to unlock your device every time you want to use it, this can be the difference in protecting your information and accounts if your smart device was to be lost or stolen.

Effective password management is crucial in safeguarding your online presence against unauthorized access and cyber threats. By understanding the importance of strong passwords, utilizing password managers, and enabling two-factor authentication, you can significantly reduce your vulnerability. The strength of your security is not just in the complexity of your passwords but in how you manage and protect them.

Reflect and Apply Questions

1. **Review Your Password Practices:** When was the last time you updated your passwords? Are there any accounts for which you've been using the same password for an extended period?

2. **Evaluate Your Password Manager Usage:** If you're already using a password manager, consider when you last explored its features fully. Are there ways you could use it more effectively to enhance your security?
3. **Assess Your 2FA Setup:** Which of your accounts currently have two-factor authentication enabled? Identify at least one important account that could benefit from added security and take steps to enable 2FA.

Chapter 3: Safe Web Browsing Practices

Navigating the internet safely is necessary in today's interconnected world. With the rise of cyber threats, protecting your personal information during web browsing is not just advisable; it's essential. This chapter guides you through practical steps and tools to ensure your online activities are secure, from understanding the basics of web security to employing advanced privacy measures.

Secure Browsing Habits

Web browsing can expose you to malware, which includes software like viruses and ransomware designed to damage your device or steal data. Phishing attacks, where attackers masquerade as trustworthy entities to steal sensitive information, are particularly common. Awareness and education on these threats can significantly mitigate risks. Understanding how these attacks are carried out—through malicious links, fake websites, or deceptive emails—empowers you to navigate the web more cautiously, recognizing and avoiding potential threats before they impact your digital life.

Navigating Public Wi-Fi Safely

Public Wi-Fi networks, though convenient, pose significant security risks. These networks are often not encrypted, allowing cybercriminals to easily intercept your data. To navigate these networks safely, always verify the network's authenticity by confirming its name with staff or signage, especially in public spaces like cafes or hotels. Avoid accessing sensitive information such as bank accounts or entering passwords while connected to public Wi-Fi. If necessary, use secure websites (HTTPS) for browsing, and consider turning off Wi-Fi when not in use to prevent automatic connections to potentially unsafe networks.

The Significance of HTTPS

HTTPS (Hyper Text Transfer Protocol Secure) ensures that the data sent between your browser and the website you're visiting is encrypted. Look for the padlock icon in your browser's address bar as a sign of a secure connection. This encryption is critical when submitting sensitive information online, such as credit card numbers or login credentials, as it protects it from being inter-cepted by unauthorized parties.

Privacy Tools and Extensions

A Virtual Private Network (VPN) encrypts your internet connection, creating a secure tunnel for your data and masking your online activities from eavesdroppers, making it an essential tool for enhancing privacy, especially on public Wi-Fi. When selecting a VPN, look for features like strong encryption stand-ards, a no-logs policy, and the availability of servers in multiple countries. This ensures the security of your data and the flexibility to access content from various locations. Implementing a VPN can significantly reduce your vulnerability to cyber threats and protect your privacy online.

Popular VPNs:

- **NordVPN**: Known for its strong security features, a large network of servers, and fast connection speeds.
- **ExpressVPN**: Offers a wide range of server locations, high-speed connections, and is praised for its ease of use.
- **CyberGhost**: Features a user-friendly interface, with a focus on accessibility and a wide array of servers tailored to specific streaming services.

Implementing Ad Blockers and Anti-Tracking Tools

Ad blockers and **anti-tracking tools** enhance your brow-sing experience by blocking intrusive ads and preventing websites from tracking your online activities. This speeds up your browsing and reduces your exposure to malicious ads that might harbor malware. Also, anti-tracking tools help protect your privacy by preventing advertisers from collecting data about your online behavior.

Popular Ad Blockers:

1. **uBlock Origin**: Highly efficient, lightweight, and cust-omizable, known for blocking ads and trackers without heavily impacting system resources.
2. **Adblock Plus**: Offers customizable filtering rules, allowing users to block most ads while enabling non-intrusive advertising to support websites.

3. **AdGuard**: Not only blocks ads across browsers but also offers protection from phishing and malicious websites, with versions for desktop and mobile devices.

Popular Anti-Tracking Tools:

1. **Privacy Badger**: Automatically learns to block invisible trackers and control outgoing data, focusing on improving privacy without much user intervention.
2. **DuckDuckGo Privacy Essentials**: Offers tracker blocking along with a privacy grade rating for websites, available as a browser extension.
3. **Ghostery**: Identifies and blocks trackers to enhance privacy and page load speeds, with detailed controls over which trackers to block.

Recognizing Malicious Websites and Downloads

Staying safe online also means being vigilant about the websites you visit and the files you download. Look for signs of credibility on websites, such as privacy policies and secure (HTTPS) connections. Be cautious of downloading files from unknown sources, as they may contain malware. Employing web security tools like web reputation services can also alert you to potentially dangerous sites before you visit them.

Safe web browsing is a critical component of your overall cybersecurity strategy. By understanding the risks, employing secure connections, protecting your privacy with VPNs, and being cautious about where you click and what you download, you can significantly reduce your vulnerability to online threats. Cyber-security is not just about the technology but also about adopting a mindset of vigilance and precaution.

Reflect and Apply Questions

1. **Assess Your Browsing Habits:** When browsing the web, do you regularly check for HTTPS or the padlock icon in your browser's address bar? How often do you connect to public Wi-Fi without using a VPN?
2. **Review Your Tools:** Do you currently use an ad blocker or anti-tracking tool? If so, how has it changed your browsing experience? If not, what concerns might you have about starting?

3. **Evaluate Your Awareness:** Reflect on a time you encountered a suspicious website or email. What red flags did you notice? How did you respond?

Chapter 4: Phishing Attacks

The digital landscape offers vast opportunities for learning, connection, and convenience. However, it also presents a parallel universe of threats, with phishing at the forefront. Phishing attacks, sophisticated in their deceit, exploit the one vulnerability that remains constant: human psychology. This chapter delves deep into the murky waters of phishing, including its various forms such as email scams, targeted spear phishing, deceptive smishing (SMS phishing), and the voice-based vishing. By understanding these threats and the psychological manipulations employed, you'll be better equipped to navigate the digital world securely.

Understanding Phishing

Phishing is a deceptive practice where attackers masquerade as trustworthy entities to trick you into revealing personal or financial information. This digital con artistry can manifest in various forms, such as emails that mimic legitimate businesses, fake websites that look real but are designed to steal your data, or urgent text messages that lead you to malicious sites. The goal is often to capture sensitive data like login credentials, credit card numbers, or social security numbers.

Types of Phishing:

- **Email Phishing**: The most common, where attackers send fraudulent emails designed to elicit personal information or financial data.
- **Spear Phishing**: A more targeted approach, where the attacker chooses specific individuals or organizations to deceive, often using personal information to increase their success rate.
- **Smishing and Vishing**: Smishing involves sending mis-leading text messages, while vishing involves phone calls. Both of these attacks aim to trick the recipient into disclosing sensitive information.

Common Tactics Used in Phishing Attacks

Attackers employ clever tactics to make their phishing attempts convincing:

- **Urgency:** Phishing messages often convey a sense of emergency, prompting you to act quickly without thinking. For example, claiming your account will be locked unless you update your information immediately.
- **Familiarity:** Using information about you, like your name or job title, to make the attack seem personalized and legitimate.
- **Impersonation:** Copying the design, logo, and language of real organizations to create emails or websites that look authentic.
- **Fear**: Instilling fear of negative consequences. Example: A message suggesting your security has been compromised, urging you to verify your credentials.
- **Authority**: Impersonating figures of authority to exploit trust. Example: An email mimicking a government agency demanding personal information.

Case Studies on Phishing Attacks

Case Study 1: **The Financial Institution Impersonation** A well-known bank's customers received emails claiming their accounts had been temporarily locked due to suspicious activities. The email directed them to a counterfeit website to verify their identity, capturing their login credentials. The aftermath included financial losses for numerous customers before the bank identified and halted the scam, enhancing their security measures and customer education programs.

Case Study 2: The CEO Fraud - An employee at a medium-sized company received an email from what appeared to be the CEO, requesting an urgent wire transfer to a specified account for a confidential deal. The employee complied, only to discover it was a phishing scam. The incident led to significant financial loss and prompted the company to implement stricter verification processes for financial transactions.

Case Study 3: The Grandparent Scam - An elderly person received an email that appeared to be from a grandchild, claiming they were in trouble and needed money immediately for bail. Moved by urgency and concern, the grandparent sent the funds via a wire transfer. It was later discovered that the grandchild was never in trouble, and the email was a phishing scam exploiting the grandparent's love and trust. This incident highlights the need for increased awareness and education on cybersecurity threats, especially among vulnerable populations.

Best Practices for Verification

To spot phishing attempts, be vigilant about certain red flags:

- **Check the Sender's Email Address:** It might look legitimate at a glance but often contains slight alterations meant to deceive, such as misspellings or additional characters.
- **Look for Language and Grammar Errors:** Phishing attempts may contain unusual phrasing, spelling mistakes, or grammatical errors not typical of professional commu-nication from genuine organizations.
- **Beware of Suspicious Links or Attachments:** Hover over links to see the actual URL before clicking. Be cautious of emails that prompt you to download attachments or click on links, especially if unsolicited.
- **Unsolicited Requests for Personal Information:** Legitimate companies rarely ask for sensitive information through email or text. You can always verify through official channels if you're unsure.

Checklist for Verifying Emails

1. **Check the Sender's Email Address**: Verify if the email address exactly matches the known address of the sender.
2. **Look for Spelling and Grammar Mistakes**: Legit-imate organizations usually send out well-written emails.
3. **Be Wary of Unsolicited Attachments**: Don't open attachments unless you're expecting them.
4. **Hover Over Links**: Before clicking, hover over any links to see where they actually lead.
5. **Verify Requests for Personal Information**: Legit-imate companies rarely ask for sensitive information via email.

6. **Contact the Sender Directly**: If unsure, contact the company or individual directly using contact information from their official website, not the email.

Protective Measures Against Phishing

To safeguard against phishing:

- **Verify the Source**: Always double-check the sender's details before responding to requests for personal info.
- **Use Security Software**: Install and maintain updated antivirus software that includes phishing protection.
- **Educate Yourself and Others**: Stay informed about the latest phishing techniques and share knowledge with friends and family.

What to Do If You Fall for a Phishing Attack

If you suspect you've responded to a phishing attempt:

1. **Change Your Passwords**: Immediately update passwords for any compromised accounts.
2. **Alert the Affected Institutions**: Contact your bank or service provider to report the incident.
3. **Report the Phishing Attempt**: Notify authorities such as the FTC in the USA or Action Fraud in the UK.

Tools and Resources for Phishing Protection

Several tools can help detect and block phishing attempts:

- **Email Filters**: Many email services include phishing filters that automatically detect and isolate suspicious emails. Services like Gmail and Outlook have built-in phishing detection systems. Gmail, for example, uses machine learning algorithms to flag suspicious emails and direct them to a Spam folder, while Outlook provides additional features like link scanning to assess the safety of embedded URLs.
- **Security Software**: Comprehensive security suites offer real-time protection against phishing sites. Norton 360 and McAfee Total Protection are comprehensive suites offering real-time protection. Norton 360 includes features like Safe Web, which warns users about malicious websites, and McAfee's WebAdvisor performs a similar function by providing safety ratings for sites and blocking risky ones before they cause harm.

- **Education Resources**: Websites like the Anti-Phishing Working Group (APWG) and CISA offer guides and alerts on phishing prevention.

Phishing attacks are a persistent threat in the digital world, but by staying vigilant, understanding their tactics, and employing protective measures, you can significantly reduce your risk of falling victim. The most effective defense against phishing is a combination of technology, education, and common sense.

Reflect and Apply Questions

1. **Review Your Experiences**: Recall a time when you encountered a potential phishing attempt. What was the giveaway?
2. **Evaluate Your Preparedness**: Considering the protec-tive measures discussed, which ones do you currently use, and which could you implement?
3. **Practice Identification**: Look at your spam folder (safely, without clicking on any links). Can you identify any phishing attempts based on the indicators mentioned?

Chapter 5: Social Media Security

In the era of digital connection, social media platforms have become central to our daily lives, facilitating interactions, business transactions, and the sharing of personal moments. Social media not only connects us with friends and family but also serves as a platform for professional networking and personal expression. However, this digital openness can make us targets for cyber-criminals. Enhancing our understanding of social media security is important for protecting our online identity and privacy. This interconnectedness brings with it a spectrum of security risks, from data breaches to personal identity theft. This chapter will navigate the complexities of social media security, offering actionable advice to protect your digital presence effectively.

Understanding Social Media Risks

Social media platforms, while providing vast opportunities for connection, also expose users to various risks. Personal information can be harvested through insecure settings, leading to identity theft, while phishing scams can trick users into revealing sensitive data. Moreover, oversharing on these platforms can jeopardize personal safety and security. Beyond identity theft and phishing scams, social media risks include cyberbullying, doxxing (publishing private information), and exposure to inappropriate content. Malicious software can also spread through social media, compromising your device and personal data.

Privacy Settings and Controls

Each social media platform offers a range of privacy settings allowing users to control who sees their content and personal information. It's important to:

- Regularly review and customize these settings to ensure that your profile and posts are visible only to intended audiences.
- Be cautious about the amount of personal information shared in your profile, as this can be mined by malicious actors.
- Dive deeper into specific privacy features, like Instagram's "Close Friends" for stories or Facebook's

profile lock, which makes your profile visible only to friends. Regular audits of friend lists and followers can further safeguard your privacy.

Being Mindful on Social-Media

Safe sharing on social media involves a strategic approach to what personal information you make public. Being mindful of location sharing is critical, as real-time updates can inadvertently reveal your whereabouts, making you vulnerable to stalking or theft. The concept of a "digital footprint" refers to the trail of data you leave online. Even seemingly innocuous details can be pieced together by cybercriminals to build a profile of you, which could be used for identity theft or other malicious purposes. It's essential to consider the long-term implications of shared content, as once something is online, it's nearly impossible to erase completely. This awareness can guide you in making informed decisions about what to share and with whom, safeguarding your online presence. The adage "think before you click" extends to what you share on social media:

- Avoid posting sensitive personal information that could be used for identity theft, such as your home address, phone number, or financial details.
- Consider the potential long-term implications of sharing certain photos or thoughts, especially regarding privacy and personal reputation.

Recognizing and Reporting Scams

Scams on social media can take many forms, from direct messages offering too-good-to-be-true deals to posts linking to malicious websites. Recognizing scams involves being alert to certain red flags, such as unexpected messages claiming you've won a prize in a contest you don't recall entering, or social media accounts pretending to be well-known celebrities or companies to solicit personal information or money. To report these scams, most social media platforms have a reporting feature accessible through the help center or security settings. This typically involves selecting the suspicious message or account, choosing the report option, and following the prompts to categorize the issue, which helps the platform take appropriate action to investigate and mitigate the scam. Recognizing these scams often involves:

- Scrutinizing unsolicited offers or friend requests from unknown individuals.
- Verifying the authenticity of sensational or alarming posts before sharing them.
- Reporting suspicious activities to the platform to help protect the wider community.

The Importance of Strong Passwords

The foundation of personal cybersecurity on social media begins with robust password practices. Strong, unique passwords act as the first line of defense against unauthorized access to your accounts. Incorporating a mix of uppercase and lowercase letters, numbers, and symbols makes your passwords difficult to crack through brute force methods. Passphrases, which are longer sequences of words or characters, offer memorable yet secure alternatives to traditional passwords. Password managers play a crucial role by not only storing your passwords securely but also generating complex passwords and auto-filling them across dif-ferent sites, significantly reducing the risk of password reuse across multiple platforms. This approach ensures that even if one account is compromised, your other accounts remain secure.

Regular Account Monitoring

Vigilance in monitoring your social media accounts goes beyond noticing odd posts or messages. It involves actively engaging with the security tools provided by social media platforms, such as setting up login alerts to notify you of new logins from unknown devices or locations. Regularly reviewing your active sessions and the devices linked to your accounts can help you spot unauthorized access quickly. Many platforms also offer a security checkup feature that guides you through reviewing your security settings and connected apps. This proactive stance aids in early detection of any suspicious activity and reinforces your overall security posture on social media.

Navigating the complexities of social media security is a dynamic and ongoing process that requires a balance between enjoying the benefits of social connectivity and safeguarding your online presence. The landscape of cyber threats is constantly evolving, making it imperative to stay informed about the latest security practices and threats. By adopting strong password habits, actively monitoring account activity, and being equipped to recognize and respond to scams, you can create a safer social media environment for yourself and those within your network. Embracing these practices ensures that you can enjoy the vast opportunities offered by social media platforms while minimizing the associated risks.

Reflect and Apply Questions

1. **Review Your Security Settings:** When was the last time you checked your privacy settings on social media?
2. **Assess Your Sharing Habits:** Are there details you've shared on social media that could be used against you?
3. **Practice Vigilance:** How would you respond if you noticed a friend's account behaving suspiciously?

Chapter 6: Protecting Sensitive Information

In the digital world, our personal and sensitive information is more exposed than ever. From social security numbers to financial data, the information we wish to keep private faces constant threats. This chapter not only introduces the concept of encryption but also expands into practical measures for securing devices, networks, and navigating the web safely, ensuring a comprehensive defense against digital vulnerabilities.

Understanding Encryption

Encryption is the first line of defense in protecting sensitive information. Simply put, encryption transforms readable data into a coded format that can only be decoded with the right key. Imagine sending a sealed letter through the mail; only the recipient, with the key (in this case, a letter opener), can read its contents. Similarly, encryption ensures that even if data is intercepted, it remains unreadable to unauthorized parties.

- **Why Use Encryption?** Encryption helps protect your personal data from cybercriminals, ensuring that your communications, files, and online transactions are secure.
- **Common Encrypted Services:** Many services we use daily, like messaging apps (WhatsApp) and online banking, use encryption to protect data in transit.

Secure Data Storage

It is important to store personal data securely. With the rise of cloud storage, understanding where and how your data is stored can seem daunting, but it doesn't have to be.

- **Choosing a Secure Cloud Storage Service:** Look for services that offer end-to-end encryption, two-factor auth-entication, and have a strong reputation for security. Examples include services like Dropbox, Google Drive, and iCloud, but it's important to review their privacy policies and security features.
- **Tips for Safe Cloud Storage Use:** Always use strong, unique passwords and enable two-factor authentication.

Be cautious about what information you store online and consider encrypting sensitive files before uploading them.

Physical Device Security

Your physical devices (laptops, smartphones, tablets, etc.) are gateways to your personal data. Securing these devices is as important as protecting the data they hold.

Best Practices for Device Security:

- **Passwords and Biometrics:** Use strong, complex pass-words and take advantage of biometric security features like fingerprint or facial recognition.
- **Software Updates:** Regularly update devices to protect against security vulnerabilities.
- **Wi-Fi Security:** Be wary of public Wi-Fi networks. Use a VPN (Virtual Private Network) for a secure connection when accessing personal information on public networks.
- **Physical Security Measures:** Consider using physical locks for laptops when in public places and never leave devices unattended.

Encryption in Action

Encrypting your data is a powerful way to safeguard it. Here's how you can apply encryption in your everyday digital activities:

- **Encrypting Smartphone Data:** Most modern smart-phones come with built-in encryption capabilities. Ensure it's enabled in your security settings to protect your data if your phone is lost or stolen.
- **Using Encryption Tools for Personal Files:** For additional security on your computer, use encryption tools like BitLocker (Windows) or FileVault (Mac) to encrypt your hard drive. Free encryption software, such as VeraCrypt, can also secure specific files and folders.
- **Email Encryption:** Use email services that offer end-to-end encryption for sensitive communications. For an extra layer of security, consider tools like ProtonMail or PGP (Pretty Good Privacy) encryption for your email.

Dealing with Sensitive Information

Sharing and handling sensitive information online requires care:

- **Safely Sharing Sensitive Information:** Avoid sharing sensitive details over unencrypted channels. If you must share personal information online, ensure the website uses HTTPS, indicating a secure, encrypted connection.

- **Recognizing Secure Websites:** Look for the padlock icon next to the website's URL, signaling that the site uses SSL (Secure Sockets Layer) encryption. This is especially important when entering financial data or personal information on a website.

Broadening Your Digital Defense: Practical Security Measures

Let's journey through the nuances of public Wi-Fi safety, fortify the security of our home networks, and spotlight the pivotal role of mobile security. These sections are designed to empower you with actionable steps, ensuring your personal information remains protected across all digital fronts.

Understanding the Risks of Public Wi-Fi

Public Wi-Fi networks, often available in cafes, airports, and hotels, provide convenient internet access but are not always secure. These networks can be a breeding ground for cyber threats like man-in-the-middle attacks, where a hacker intercepts the communication between your device and the internet. Unencrypted networks allow hackers to easily access unsecured devices on the same network. To use public Wi-Fi safely, always verify the network name with staff to avoid rogue hotspots, use a VPN (Virtual Private Network) to encrypt your internet traffic, and avoid accessing sensitive information like banking sites while connected.

Securing Your Home Network Made Simple

Securing your home network might sound complex, but it's like locking your doors at night—it's about keeping unwanted guests out. Here's how to do it in simple steps:

1. **Change Default Settings:** Your Wi-Fi router comes with a preset username and password that hackers can easily guess. Think of it as leaving your key under the mat. Changing these to something only you know is like up-grading to a deadbolt.
2. **Use Strong Encryption:** Encryption scrambles the information sent over your network so it can't be easily

read by outsiders. Most routers offer different types of encryption, and WPA3 is the newest, like having the latest security system for your home. Make sure it's turned on in your router's settings.

3. **Keep Your Router Updated:** Just like apps on your phone, your router needs updates to fix security holes. Check the manufacturer's website to learn how to update your router, similar to changing the batteries in your smoke detector to keep it functioning.
4. **Know Who's Connected:** Keep an eye on which devices are connected to your network. It's like knowing who's in your house. If you see something unfamiliar, it might be time to change your Wi-Fi password.
5. **Turn Off WPS:** Wi-Fi Protected Setup (WPS) is a feature that lets you connect devices with the push of a button. However, it can also make it easier for hackers to connect. Turning it off is like not hiding a spare key outside.

By following these steps, you're not just securing your Wi-Fi; you're protecting everything and everyone connected to it.

Spotlight on Mobile Security

Our smartphones are like mini-computers we carry in our pockets, filled with personal information. Here's how to keep them secure:

1. **App Permissions:** Be cautious about which apps you allow access to your information. Think of app permissions like giving someone a key to your house. Only give it to those you trust.
2. **Avoid Sideloading Apps:** Sideloading is when you download apps from places other than the official app store. This can be risky, like buying a used car without knowing its history. Stick to the official app stores for safer downloads.
3. **Secure Your Device:** Use a strong passcode or biometric lock (like your fingerprint or face recognition) on your phone. It's the first line of defense if your phone gets lost or stolen, acting as a lock on your front door.
4. **Update Regularly:** Keep your phone's operating system and apps up to date. These updates often include fixes for

security vulnerabilities, akin to fixing a broken window to keep burglars out.

5. **Be Wary of Public Wi-Fi:** Just like with your home network, be cautious when using public Wi-Fi. If you must, use a VPN to protect your data, like having a security escort in a crowded place.

By taking these simple steps, you can significantly reduce the risk of personal information being compromised, keeping your digital life as secure as your physical one.

Protecting sensitive information may seem daunting, but by taking proactive steps and utilizing encryption, you can significantly mitigate risks. This chapter has equipped you with the knowledge and tools to build a comprehensive digital defense strategy. By staying informed, vigilant, and proactive in implementing these practices, you can substantially mitigate risks and ensure a safer digital experience. In the realm of cybersecurity, your actions today define your security tomorrow.

Reflect and Apply Questions

1. **Review Encryption Practices:** Reflect on the encryption tools and services you currently use. Are your devices and sensitive files encrypted to protect against unauthorized access?
2. **Evaluate Your Vigilance:** Consider how you verify the security of websites and emails before sharing personal information. How can you improve your practices to ensure you're always using secure platforms?
3. **Explore Email Security:** Reflect on your email security practices. Have you considered using encrypted email services or encryption tools for sensitive communications? What changes can you make to enhance your email security?
4. **Network Safety Measures:** Consider your home network and mobile device security. How can you apply the discussed measures to fortify your digital fortress against external threats?
5. **Public Wi-Fi Awareness:** How does your current use of public Wi-Fi reflect an understanding of its risks? What changes can you implement to ensure safer browsing practices?

Chapter 7: Staying Informed and Up-to-Date

In our ever-changing digital landscape, cybersecurity threats evolve at an astonishing pace. It's vital for individuals, not just professionals, to stay informed and adapt to these changes. This chapter aims to empower you with the knowledge and tools necessary to keep abreast of the latest developments in cybersecurity, ensuring you and your loved ones remain protected in this digital era.

The Importance of Continuous Learning

Cybersecurity is a field that never stands still. The techniques that hackers use today might be obsolete tomorrow, and new vulnerabilities are discovered daily. Continuous learning is not just beneficial; it's essential for anyone who wants to safeguard their personal information against these evolving threats. By staying informed, you're building a resilient defense against potential cyber-attacks, much like how a flu vaccine needs to be updated every year to combat new strains.

Keeping Up with Cybersecurity News

Staying updated with cybersecurity news is easier than you might think. Many reputable websites and news outlets offer the latest information on threats and protection strategies. Start with well-known sources like Wired's Security section, The Hacker News, and Krebs on Security. These sites break down complex information into more understandable pieces, ideal for non-technical readers. Setting up Google Alerts for cybersecurity topics can also bring the latest news straight to your inbox.

To efficiently keep up with cybersecurity news, utilizing tools like Feedly can significantly streamline the process. Feedly is a news aggregator application for various web browsers and mobile devices running iOS and Android. It compiles news feeds from a variety of online sources for the user to customize and share with others. By setting up a Feedly account, you can subscribe to feeds from your preferred cybersecurity news websites, blogs, and other sources. This allows you to have a centralized hub of information, making it easier to stay informed about the latest cybersecurity trends, threats, and best practices without having to visit each site individually. Feedly's user-friendly interface and customization options make it an invaluable tool for anyone looking to enhance their knowledge and awareness in the field of cybersecurity.

Leveraging Social-Media for Learning

Social media platforms are not just for socializing; they can be powerful learning tools. LinkedIn and X (Twitter), for example, host a plethora of cybersecurity experts who regularly share insights, news, and tips. Following these professionals can provide you with a steady stream of valuable information. Look for individuals and organizations with proven expertise, such as Bruce Schneier or the Electronic Frontier Foundation, to start building a feed filled with high-quality cybersecurity content.

Participating in Online Forums and Communities

Online forums and communities offer a treasure trove of information, where both novices and experts gather to discuss the latest in cybersecurity. Websites like Reddit have dedicated subreddits where members share news, ask questions, and offer advice. Stack Exchange also hosts a cybersecurity section, where you can find answers to technical questions or participate in discussions. Facebook and LinkedIn both have many different groups created for learning and staying up to date in the industry. Engaging in these communities can enhance your understanding and provide practical advice for protecting your digital life.

Educational Resources and Courses

The internet is rich with resources for learning about cybersecurity. Websites like Coursera, edX, and Udemy offer courses taught by industry professionals. These platforms provide both free and paid courses, covering basics for beginners to advanced topics for those looking to deepen their knowledge. Additionally, many universities and institutions offer open course-ware in cybersecurity, allowing free access to course materials.

In my opinion, one heavily underrated resource is YouTube. You can find many different videos for a vast amount of Information Technology topics including cybersecurity. There are even bite sized videos to help understand a particular concept like encryption or securing your home network. If you have interest in obtaining an industry certification, check there first for course material. I would recommend watching videos by industry influencers like David Bombal or Network Chuck.

Implementing What You Learn

It's not just about gathering information; it's about putting that knowledge into action. Start by applying basic cybersecurity measures in your digital life, such as updating software regularly, using strong passwords, and enabling two-factor authentication on your accounts. Practice safe browsing habits, be skeptical of unsolicited emails, and educate friends and family on the importance of cybersecurity.

Websites like Hack the Box and Try Hack Me allow you to implement the topics you learn. These are game like environments where you pick certain topics like cybersecurity or ethical hacking and follow through the material. You learn a concept and then put it into action on a Virtual Machine (VM) right on the site. These are excellent for hands-on learners

The Role of Cybersecurity Tools

Familiarize yourself with basic cybersecurity tools that can help monitor and protect your digital presence. Antivirus software, firewalls, and VPNs are fundamental tools that can provide a significant layer of security. Additionally, consider using password managers to securely store and manage your passwords. Security tools that alert you to data breaches involving your personal information are also great to have.

As we conclude our journey through chapter 7, remember that cybersecurity is a continuous process. The digital world evolves, and so do the threats within it. By taking proactive steps to educate yourself, stay updated on the latest news, and implement what you learn, you're not just protecting your digital footprint—you're enhancing your overall digital literacy in a way that benefits every aspect of your online life.

Reflect and Apply Questions

1. **Explore Learning Resources:** Which online platform will you explore first to find a cybersecurity course that interests you? Consider setting a goal to complete one course within the next six months.
2. **Evaluate Your Cybersecurity Posture:** Reflect on your current cybersecurity practices. Which area do you

feel needs the most improvement, and what steps will you take to address it?

3. **Engage with the Community:** How can participating in online forums or cybersecurity communities enhance your understanding and application of cybersecurity practices? Identify one community you will join and contribute to.
4. **Leverage Tools for Security:** Identify one cyber-security tool you haven't used before but are interested in trying out. Research how it could improve your personal or professional digital security.

Conclusion

Throughout this book, we've navigated the essential practices of cybersecurity, from understanding the basics of digital threats to implementing practical measures for safeguarding our infor-mation. We've explored encryption, safe browsing habits, the importance of secure passwords, and the critical nature of staying informed in the face of rapidly evolving cyber threats. This journey has equipped you with the tools to protect your digital life and has hopefully sparked a desire to learn more.

For those who are inspired to deepen their understanding of cybersecurity or explore the broader landscape of Information Technology, I encourage you to connect with me. Further your education and career in this dynamic field by diving into my book, "Breaking Into IT: Navigating and Thriving in the Tech Industry." It's designed not just to inform but to inspire action and continued growth in the ever-changing tech landscape.

Take this knowledge, apply it, and never stop learning. The digital world is vast and full of opportunities for those who are prepared. Let's continue this journey together, building safer and more secure digital environments for ourselves and those around us.

Want more?

Check out my debut book, "Breaking Into It: Your Roadmap to Success in the Tech Industry." You can also visit my website www.unitethegrid.com. We are the GRID, a collective of IT professionals that aim to provide all of the resources you need to excel in your IT career in one location. We offer books, study sheets, mentoring, coaching, recommendations, and more to come! Come check us out and accelerate your path to a rewarding and successful career in IT! If you found this book helpful along your journey, I encourage you to return to the platform where you obtained the book to leave a good review. This will help others find the book, allowing it to help a wider audience.

Appendix A: Further Reading and Resources

StaySafeOnline

- **Website:** https://staysafeonline.org
- **Description:** A resource provided by the National Cyber Security Alliance, offering cybersecurity awareness and education for consumers, small and medium-sized businesses, and schools. Readers can find tips on protecting personal information and advice on improving online safety practices.

Cybersecurity & Infrastructure Security Agency (CISA)

- **Website:** https://www.cisa.gov
- **Description:** The United States government agency dedicated to enhancing the nation's cybersecurity and infrastructure protection. CISA's website provides alerts on current cybersecurity threats, vulnerabilities, and offers resources on how to improve cybersecurity posture.

Krebs on Security

- **Website:** https://krebsonsecurity.com
- **Description:** A blog by journalist Brian Krebs that covers in-depth cybersecurity topics, including the latest cyber threats, security news, and investigative reports on cybercrime. This blog is a valuable resource for those looking to dive deeper into cybersecurity issues and trends.

Appendix B: Citations

Mogg, T. (2019, April 21). Online passwords: Research confirms millions of people are using 123456. *Digital Trends.* https://www.digitaltrends.com/computing/online-passwords-research-confirms-millions-are-using-123456/

Made in the USA
Monee, IL
04 November 2026

67384412R00026